PICTURING THE CHEMICAL ELEMENTS

REPRESENTING EACH OF THE ELEMENTS WITH A RELATED PICTURE, INCLUDING: MAPS, STAMPS, COAT OF ARMS, SCIENTISTS

ORGANIZED AND WRITTEN BY CATHERINE JAIME

All pictures are either in the public domain or used with permission.

Creative Learning Connection
www.CreativeLearningConnection.com
www.CatherineJaime.com
Copyright © 2016 by Catherine Jaime

Introduction

A visual tour of the periodic table. Each element is shown with a picture of some kind – based on the element itself, how it is used, where it was found, who discovered it, or who it was named for.

This book can be used by itself, or as a complement to "Naming the Chemical Elements."

Happy Learning
Cathy and Crew

Dmitry Ivanovich Mendeleev (creator of the Periodic Table)
Portrait by Ilya Repin, 1885

NGC 604, ionized Hydrogen in the Triangulum Galaxy. Hydrogen was discovered in 1766.
#1, Hydrogen, H

Helium is used in dirigibles.
#2, Helium, He

Johan August Arfwedson, Swedish chemist, is credited with the discovery of Lithium in 1817.
#3, Lithium, Li

Louis-Nicolas Vauquelin, French chemist, analyzed Beryllium in 1798.
#4, Beryllium, Be

Discovered in France in 1808.
#5, Boron, B

Antoine-Laurent de Lavoisier, French chemist, recognized Carbon in 1789.
#6, Carbon, C

Isolated in Scotland in 1772.
#7, Nitrogen, N

American Robert H. Goddard with a liquid oxygen-gasoline rocket in 1926.
#8, Oxygen, O

Fluorite being used as an additive to lower the melting point of metals during smelting.
#9, Fluorine, F

Vanity Fair's caricature of Sir William Ramsay, Scottish chemist who discovered the Noble Gases and received the Nobel Prize in Chemistry in 1904.
#10, Neon, N

Close Up of Sodium Silicate Isolated by Cornish chemist in 1807.
#11, Sodium, S

Joseph Black, Scottish chemist who discovered Magnesium in 1755.
#12, Magnesium, Mg

First isolated by a Danish chemist in 1825.
#13, Aluminium, Al

Predicted by de Lavoisier in 1787.
#14, Silicon, Si

The Alchemist Discovering Phosphorus
Painting by Joseph Wright in 1771.
#15, Phosphorus, P

Discovered by the Chinese more than 4,000 years ago.
#16, Sulfur, S

Discovered by a Swedish chemist in 1774.
#17, Chlorine, Cl

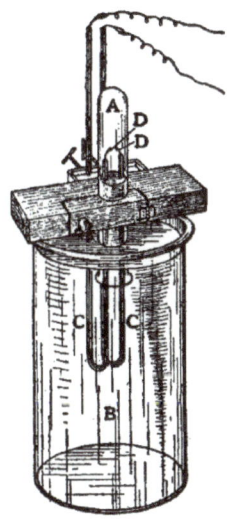

Lord Rayleigh's method of isolating Argon
#18, Argon, Ar

Cornish chemist, Humphry Davy, discovered Potassium in 1807 and Calcium in 1808.
#19, Potassium, K
#20, Calcium, Ca

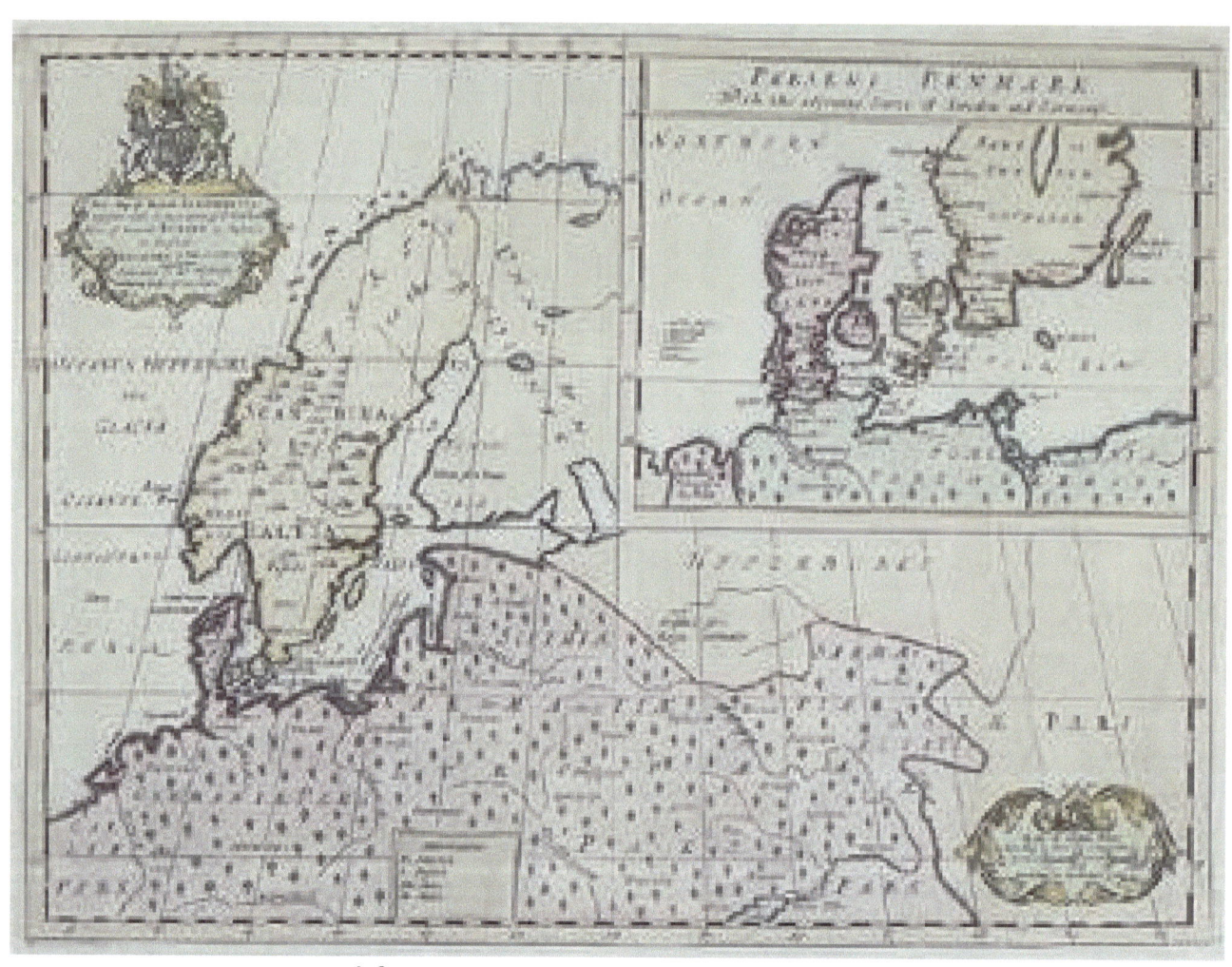

Named for Scandinavia after it was discovered by a Swedish chemist in 1879.
#21, Scandium, Sc

Discovered by a British mineralogist in 1791.
#22, Titanium, Ti

Discovered by Andrés Manuel del Río, a Spanish–Mexican scientist and naturalist, in 1801.
#23, Vanadium, V

Discovered by a French chemist in 1797.
#24, Chromium, Cr

Johan Gottlieb Gahn is given credit for first isolating pure Manganese in 1774.
#25, Maganese, Mn

An Iron Forge
Painting by Joseph Wright of Derby in 1773.
#26, Iron, Fe

Cobalt Blue Glassware
Cobalt has been used to color glass for at least 3500 years.
#27, Cobalt, Co

Ravensthorpe Nickel Operation Plant in Australia where nickel has been mined since 2002.
#28, Nickel, Ni

Bingham Copper Mine in Utah where copper has been mined since 1863.
#29, Copper, Cu

Andreas Sigismund Marggraf is given credit for first isolating pure zinc in 1746.
#30, Zinc, Zn

French chemist François Lecoq de Boisbaudran discovered Gallium in 1875.
#31, Gallium, Ga

Named for Germany after discovered by a German chemist in 1886.
#32, Germanium, Ge

Satirical cartoon by Honoré Daumier of a chemist giving a public demonstration of Arsenic in 1841.
#33, Arsenic, As

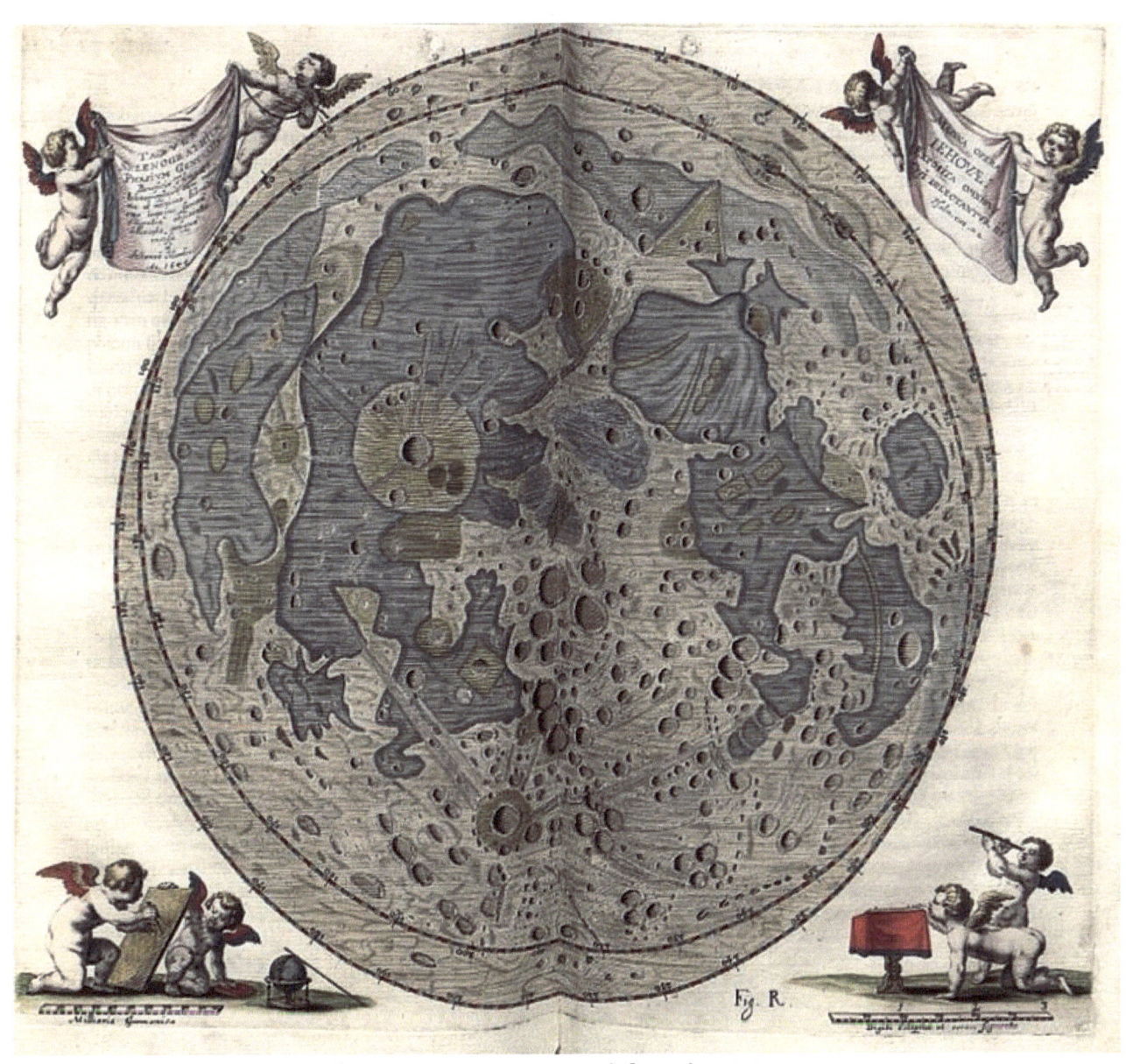

*Element was named for the Moon.
Map of the Moon by Johannes Hevelius
from his 1647 "Selenographia."*
#34, Selenium, Se

In 1826 Antoine Jérôme Balard (French chemist) was one of two scientists to independently discover Bromine.
#35, Bromine, Br

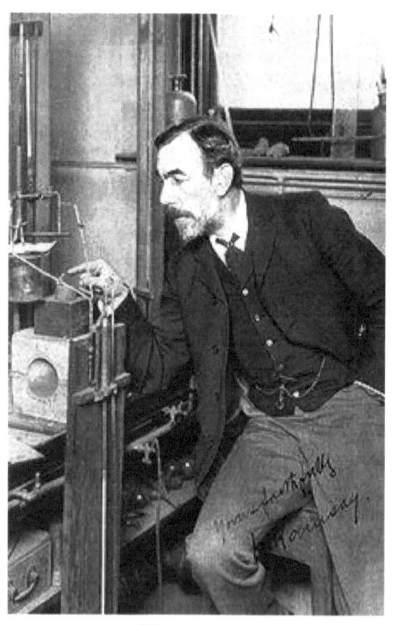

Sir William Ramsay discovered Krypton in 1898.
#36, Krypton, Kr

Gustav Kirchhoff (left) and Robert Bunsen (right) discovered Rubidium in 1861.
#37, Rubidium, Rb

Flame test for Strontium Strontium was discovered in Scotland in 1790.
#38, Strontium, Sr

One of several rare elements discovered near Ytterby, Sweden, this one in 1787.
#39, Yttrium, Y

Martin Heinrich Klaproth (German chemist) discovered Zirconium (and Cerium and Uranium).
#40, Zirconium, Zr (1789)

Charles Hatchett (English chemist) discovered Niobium in 1801.
#41, Niobium, Nb

Climax Molybdenum Mine in Colorado, c.1924
#42, Molybdenum, Mo

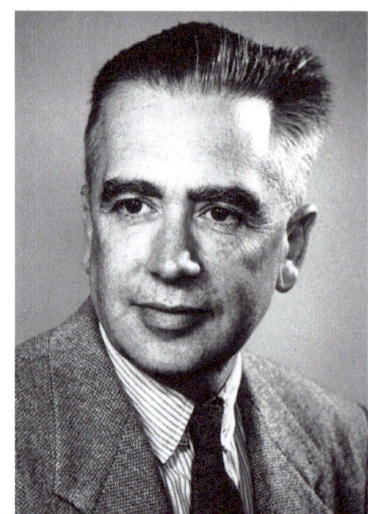

Emilio Segrè, Italian physicist and one of the discoverers of Technetium (and Astatine).
#43, Technetium, Tc

Gottfried Wilhelm Osann (German chemist and physicist) and Jöns Jacob Berzelius (Swedish chemist) discovered Ruthenium in 1828.
#44, Ruthenium, Ru

William Hyde Wollaston (English chemist and physicist) discovered Rhodium and Palladium in 1803.
#45, Rhodium, Rh
#46 Palladium, Pd

Silver mining and processing in Kutná Hora, Central Europe, 1490s.
#47, Silver, Ag

German chemists Friedrich Stromeyer (shown) and Karl Samuel Leberecht Hermann both discovered Cadmium in 1817.
#48, Cadmium, Cd

German chemist Ferdinand Reich co-discovered Indium in 1863.
#49, Indium, In

Washing tin in the mines at Stanthorpe, Australia in 1872.
#50, Tin, Sn

Used in Egypt at least 5,000 years ago.
#51, Antimony, Sb

In 1782 Franz-Joseph Müller von Reichenstein (Austrian mineralogist) discovered Tellurium.
#52 Tellurium, Te

French chemist and physicist Joseph Louis Gay-Lussac recognized Iodine as a new element in 1811.
#53, Iodine, I

Discovered by the Scottish chemist William Ramsay (above) and English chemist Morris Travers in 1898.
#54, Xenon, X

Robert Bunsen (center) and Gustav Kirchhoff (left) discovered Rubidium in 1861.
#55, Caesium, Cs

Discovered by a Swedish chemist in 1772.
#56, Barium, Ba

Carl Gustaf Mosander, Swedish chemist, who discovered Lanthanum in 1838.
#57, Lanthanum, La

Named for the asteroid Ceres, which had been named after the Roman goddess of agriculture.
#58, Cerium, Ce (1803)

Carl Auer von Welsbach (Austrian scientist and inventor) discovered both elements in 1885.
#59, Praseodymium, Pr
#60, Neodymium, Nd

Named for Prometheus, a Titan from Greek mythology, after discovered in the U.S. in the 1940s.
#61, Promethium, Pm

French chemist François Lecoq de Boisbaudran also discovered Samarium.
#62, Samarium, Sm

Europium was named for Europe after it was isolated a French chemist in 1901.
#63, Europium, Eu

Named for Johan Gadolin,
a Finnish chemist, physicist
and mineralogist,
after it was discovered in 1880.
#64, Gadolinium, Gd

Carl Gustaf Mosander, Swedish
chemist who discovered terbium
and erbium in 1842.
#65, Terbium, Tb

French chemist François
Lecoq de Boisbaudran also
discovered Dysprosium.
#66, Dysprosium, Dy

Swiss chemist Per Teodor Cleve
discovered Holmium in 1878.
#67, Holmium, Ho

Discovered in 1874 by Carl Gustaf Mosander.
#68, Erbium, Er

Named after Thule, a mythical region in Scandinavia, after being isolated by a Swedish chemist in 1879.
#69, Thulium, Tm

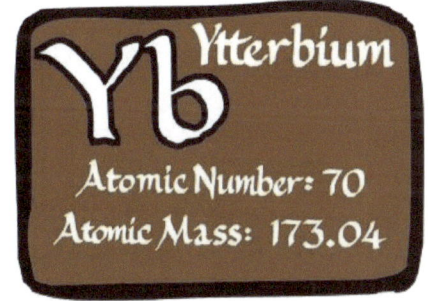

One of several rare elements discovered near Ytterby, Sweden, this one in 1787.
#70, Ytterbium, Yb

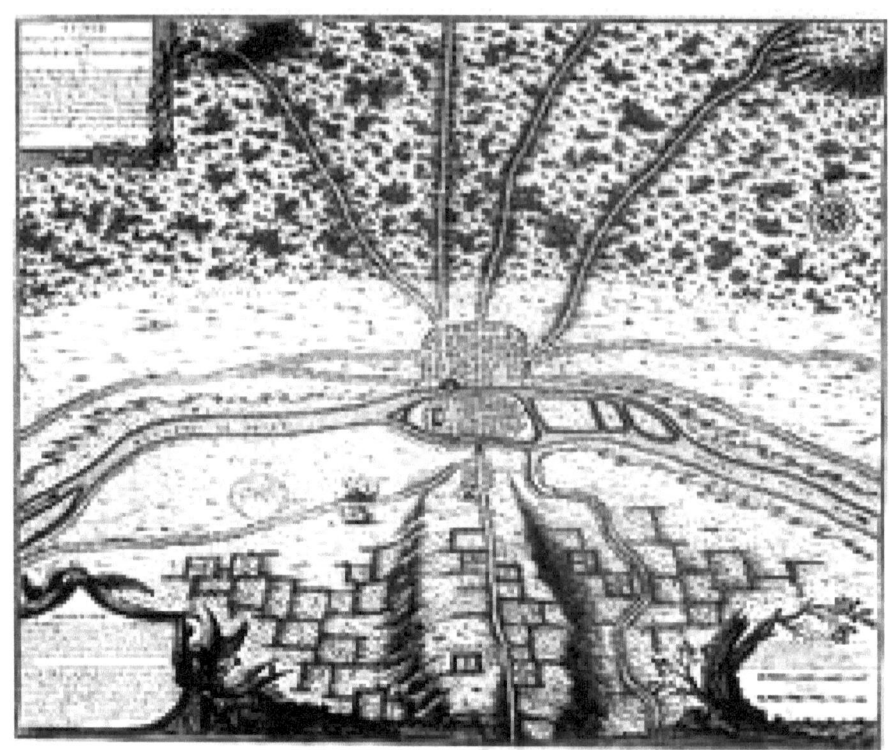

Named for Lutetia, Latin for Paris in the Roman era, after it was discovered in 1906.

#71, Lutetium, Lu

Named after Hafnia. Latin for Copenhagen, where it was discovered in 1922.

#72, Hafnium, Hf

Anders Gustaf Ekeberg,
a Swedish chemist, who
discovered Tantalum in 1802.
#73, Tantalum, Ta

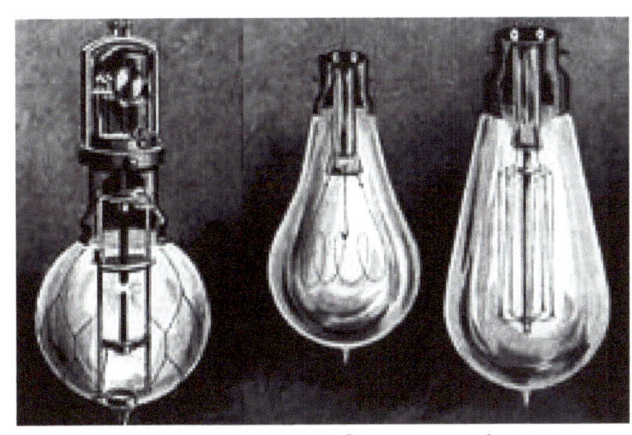

Tungsten is often used in
incandescent light bulb filaments.
Tungsten was discovered
by a Swedish chemist.
#74, Tungsten, W

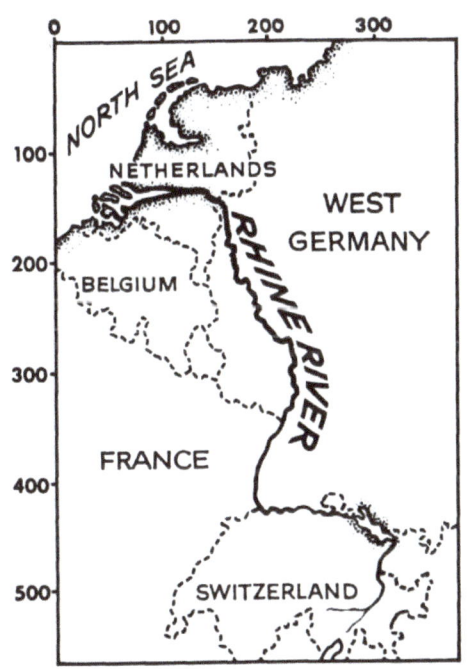

Named after the Rhine River
after it was discovered in 1908.
#75, Rhenium, Re

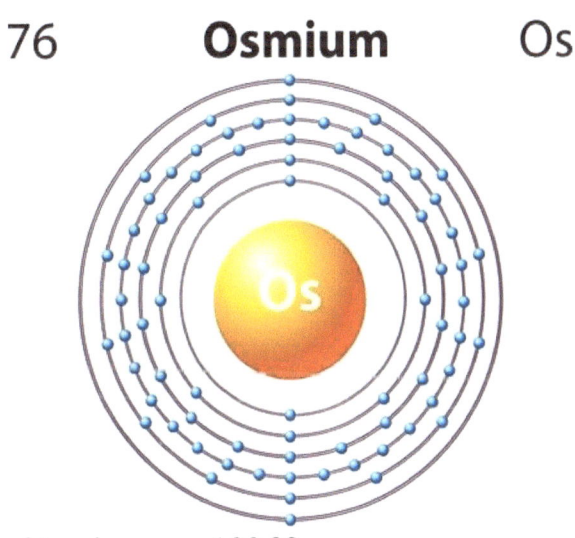

Atomic mass: 190.23
Electron configuration: 2, 8, 18, 32, 14, 2

Discovered and isolated in
1803 by an English chemist.
#76, Osmium, Os

Named for Iris, from mythology, after it was discovered and isolated in 1803 by an English chemist.
#77, Iridium, Ir

Antonio de Ulloa (Spanish astronomer) is credited with the discovery of Platinum in 1735.
#78, Platinum, Pt

Gold artifacts have been discovered dating back 6,000 years.
#79, Gold, Au

The Mercury Atmosphere and Surface Composition Spectrometer (MASCS) instrument aboard NASA's MESSENGER spacecraft took this photograph in 2015.
#80, Mercury, Hg

Sir William Crookes (English chemist) discovered Thallium in 1861.
#81, Thallium, Tl

Lead mining in the upper Mississippi River region in the United States in 1865. **#82, Lead, Pb**

Discovered in 1753 by a French chemist.
#83, Bismuth, Bi

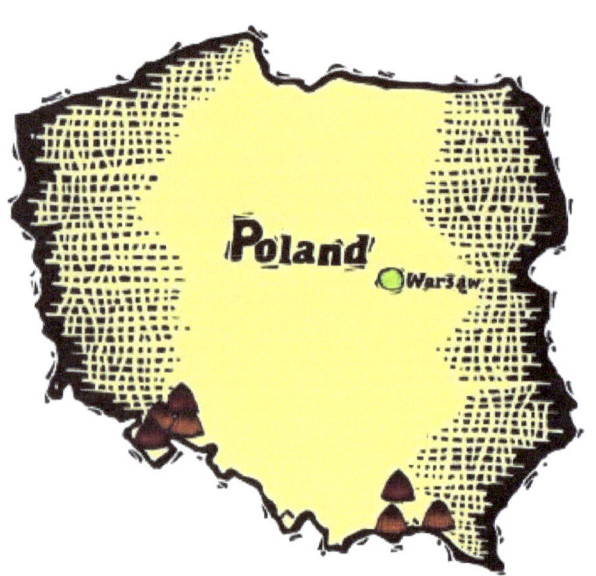

The Curies discovered Polonium in 1898 and named it for Marie's home country, Poland.
#84, Polonium, Po

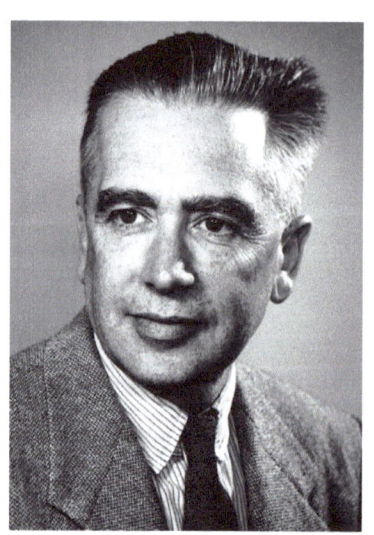

Emilio Segrè, Italian physicist and one of the discoverers of Astatine (and Technetium).
#85, Astatine, At (1940)

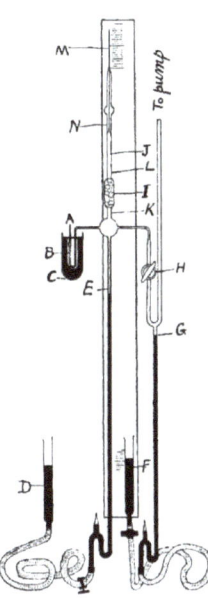

Apparatus used by Ramsay and Whytlaw-Gray to isolate Radon in London in 1910.
#86, Radon, Rn

Marguerite Perey (a French physicist and student of Marie Currie) discovered Francium in 1939.
#87, Francium, Fr

Marie and Pierre Curie experimenting with Radium in 1898.
#88, Radium, Ra

Discovered by a German organic chemist in 1902.
#89, Actinium, Ac

Named for Thor, of Norse mythology, after being discovered by a Swedish chemist in 1829.
#90, Thorium, Th

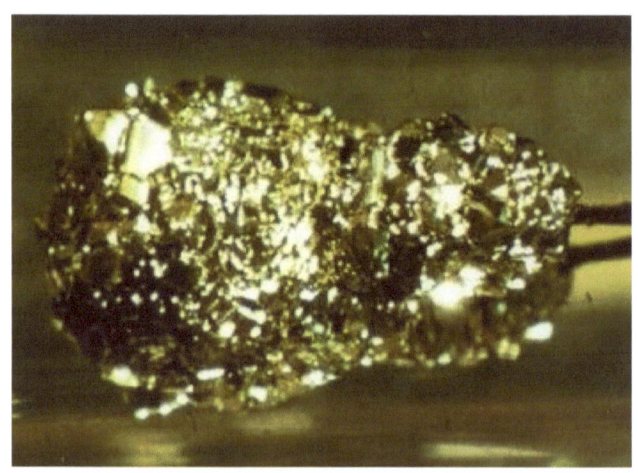

Discovered in 1900 by an English chemist and physicist.
#91, Protactinium, Pa

Named for the planet Uranus (photo taken by Voyager 2).
#92, Uranium, U

Named for the planet Neptune. Discovered by American physicists in 1940.
#93, Neptunium, Np

Named for the (then) planet Pluto. Discovered by American chemists and a physicist in 1940-1941.
#94, Plutonium, Pu

Americium was one of the Elements detected in the fallout from the Ivy Mike nuclear test the Americans conducted in 1952 (above the Pacific Ocean).
#95, Americium, Am

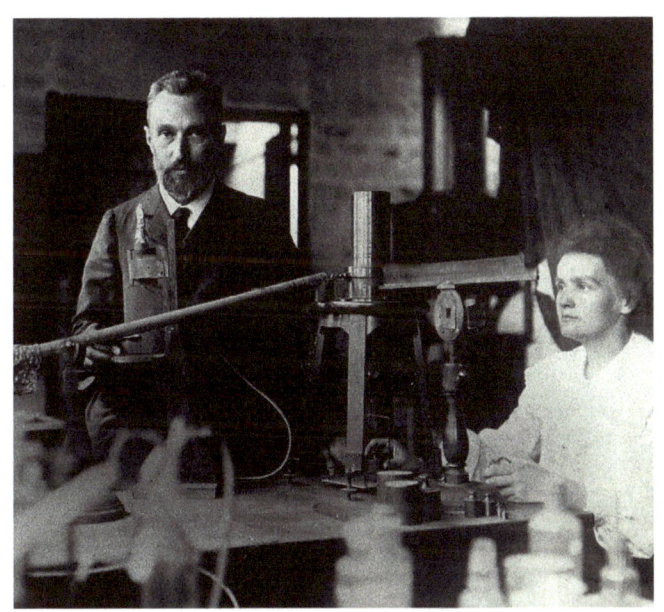

Named for Pierre and Marie Curie when it was discovered in 1944.
#96, Curium, Cm

One of several elements discovered at the University of California in Berkeley in 1949.
#97, Berkelium, Bk

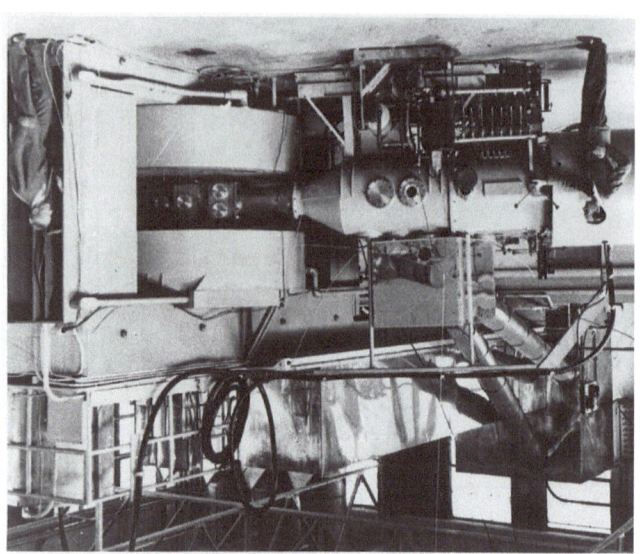

The 60-inch-diameter cyclotron used to first synthesize Californium in 1950.
#98, Californium, Cf

Named for Albert Einstein after it was discovered in 1952.
#99 Einsteinium, Es

Named after Enrico Fermi, an Italian physicist who created the world's first nuclear reactor – in Chicago, IL.
#100, Fermium, Fm

Named for (Russian chemist) Dmitri Mendeleev,
the "Father of the Periodic Table" after it was discovered in 1955.
#101, Mendelevium, Md

Reihen	Gruppe I. — R^2O	Gruppe II. — RO	Gruppe III. — R^2O^3	Gruppe IV. RH^4 RO^2	Gruppe V. RH^3 R^2O^5	Gruppe VI. RH^2 RO^3	Gruppe VII. RH R^2O^7	Gruppe VIII. — RO^4
1	H=1							
2	Li=7	Be=9.4	B=11	C=12	N=14	O=16	F=19	
3	Na=23	Mg=24	Al=27.3	Si=28	P=31	S=32	Cl=35.5	
4	K=39	Ca=40	—=44	Ti=48	V=51	Cr=52	Mn=55	Fe=56, Co=59, Ni=59, Cu=63.
5	(Cu=63)	Zn=65	—=68	—=72	As=75	Se=78	Br=80	
6	Rb=85	Sr=87	?Yt=88	Zr=90	Nb=94	Mo=96	—=100	Ru=104, Rh=104, Pd=106, Ag=108.
7	(Ag=108)	Cd=112	In=113	Sn=118	Sb=122	Te=125	J=127	
8	Cs=133	Ba=137	?Di=138	?Ce=140	—	—	—	— — — —
9	(—)	—	—	—	—	—	—	
10	—	—	?Er=178	?La=180	Ta=182	W=184	—	Os=195, Ir=197, Pt=198, Au=199.
11	(Au=199)	Hg=200	Tl=204	Pb=207	Bi=208	—	—	
12	—	—	—	Th=231	—	U=240	—	— — — —

Dmitri Mendeleev's 1871 Periodic Table,
with room for undiscovered elements.

Named for Alfred Nobel (Swedish chemist) after it was discovered in 1966.
#102, Nobelium, No

The element was named after Ernest Lawrence (the American developer of the cyclotron) after it was discovered between 1961 and 1971.
#103, Lawrencium, Lr

Named for Ernest Rutherford (British physicist) after it was discovered in 1964.
#104, Rutherfordium, Rf

Named for the town in Russia (Dubna) where it was first produced in 1968.
#105, Dubnium, Db

Named after Glenn T. Seaborg after it was discovered in 1964.
#106, Seaboium, Sg

Named after Niels Bohr after it was discovered in 1981.
#107, Bohrium, Bh

Named for the Latin name for Hesse (Germany) where it was discovered in 1984.
#108, Hassium, Hs

Named for Lise Meitner (Austrian physicist) after it was discovered in 1982.
#109, Meitnerium, Mt

Created near
Darmstadt, Germany in 1994.
#110, Darmstadtium, Ds

Named for German physicist
Wilhelm Conrad Röntgen
after it was synthesized in 1994.
#111, Roentgenium, Rg

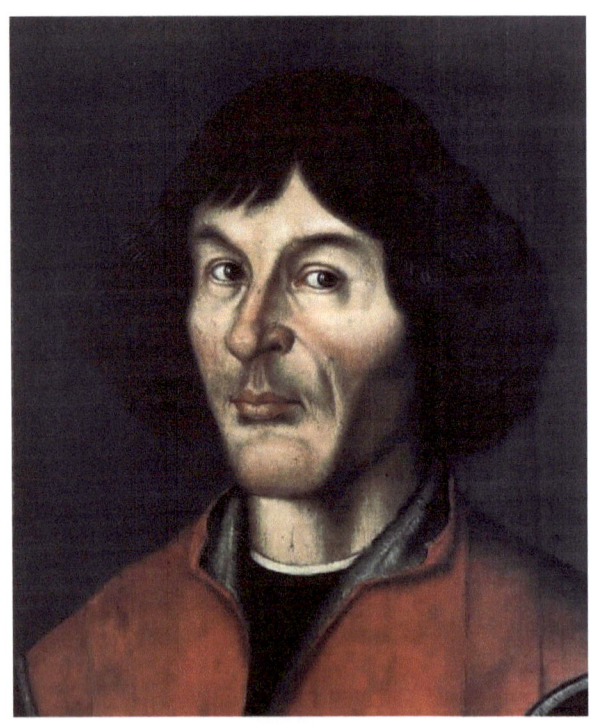

Named for Nicolaus Copernicus
after it was created in 1996.
#112, Copernicium, Cn

Temporary name for element
synthesized in Dubna, Russia
in the early 2000s.
#113, Ununtrium, Uut

Named for Georgy Flyorov, founder of the Flerov Laboratory of Nuclear Reactions in Dubna, Russia, where it was discovered in 1998.
#114, Flerovium, Fl

Temporary name for element synthesized in Dubna, Russia in the early 2000s.
#115, Ununpentium, Uup

Synthesized in 2000 in Dubna, Russia and at the Lawrence Livermore National Laboratory, in Livermore, CA (named for Robert Livermore).
#116, Livermorium, Lv

Temporary names for elements synthesized in Dubna, Russia in the early 2000s.
#117, Ununseptium, Uus
#118, Ununoctium, Uuo

www.ingramcontent.com/pod-product-compliance
Lightning Source LLC
Chambersburg PA
CBHW040753200526
45159CB00025B/2085